CAMBRIDGE PRIMARY MATHEMATICS

MODULE 9
BOOK

Roy Edwards
Mary Edwards
Alan Ward

3-95

C000145971

Cambridge University Press
Cambridge
New York   Port Chester
Melbourne   Sydney

Published by the Press Syndicate of the University of Cambridge
The Pitt Building, Trumpington Street, Cambridge CB2 1RP
40 West 20th Street, New York, NY 10011-4211, USA
10 Stamford Road, Oakleigh, Melbourne 3166, Australia

First published 1990
Reprinted 1991

Printed in Great Britain by Scotprint Ltd, Musselburgh

British Library cataloguing in publication data

Edwards, Roy
Cambridge primary mathematics.
Module 6
Bk. 1
1. Mathematics
I. Title   II. Edwards, Mary   III. Ward, Alan
510

ISBN 0 521 35824 8

The authors and publishers would like to thank the many schools
and individuals who have commented on draft material for this
course. In particular, they would like to thank Anita Straker for
her contribution to the suggestions for work with computers,
Norma Anderson, Ronalyn Hargreaves (Hyndburn Ethnic
Minority Support Service) and John Hyland Advisory Teacher
in Tameside. They would also like to thank British Rail,
Thomas Cook and The Post Office for help and information.

Photographs are reproduced courtesy of:
front cover, p56 ZEFA, pp4-5, 8 Allsport; pp6-7 by courtesy
of the Austrian National Tourist Office; p35 the Swiss National
Tourist Office; p45 Frank Lane Picture Agency; p66 The British
Sundial Society; pp67, 68 reproduced courtesy of the Trustees
of the British Museum; p70 Black Forest Tourist Office; p81 Gérard
Loez/NHPA; p82 Anthony Bannister/NHPA; p83 Eric and David
Hosking; p84 Copyright reserved; reproduced by gracious
permission of Her Majesty The Queen; pp85-9 Archbishop of
York's Junior School; p96 Record Potain Ltd.

All other photographs by Graham Portlock.

'I am a sundial' reprinted by permission of the Peters Fraser
& Dunlop Group Ltd.

The mathematical apparatus was kindly supplied by E J Arnold.
The coins were loaned by Granta Stamp and Coin Shop, Cambridge.

Designed by Chris McLeod

Illustrations by John Bendall Brunello
Diagrams by DP Press
Children's illustrations by Sarah Middleton, Robert Gilfillan,
and pupils of Scotforth C of E Primary School, Lancaster.

DP

Contents

Find the total number of people watching these events.

1	gymnastics 3183		2	cycling 1247
	swimming + 628			boxing + 576
	total ____			total ____

3 rowing 2045 and volleyball 875

4 hockey 1183 and judo 759

Find the scores. The corner numbers add up to the total.

5	6	7

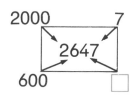

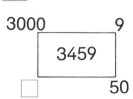

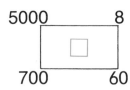

Finish the patterns.

8	145 + 99 = ☐	9	263 + 99 = ☐
	245 + 99 = ☐		363 + 99 = ☐
	345 + 99 = ☐		463 + 99 = ☐
	445 + 99 = ☐		563 + 99 = ☐

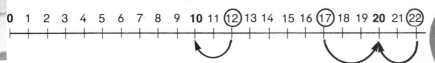

These numbers are rounded to the nearest 10.
12 is rounded to 10.
17 is rounded to 20.

10 Round these numbers to the nearest 10.

18	14	21	36	82
97	123	151	112	179

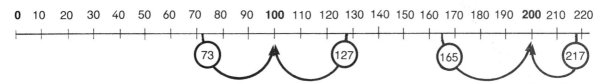

11 Round these numbers to the nearest 100.

93	215	386	435	601
317	529	842	798	560

These numbers are rounded to the nearest 100.

Complete these magic squares. Their magic number is 34.

12

= 34

16	3	2	13	= 34
5	10	☐	8	= 34
9	☐	7	12	= 34
4	15	14	☐	= 34

‖ ‖ ‖ ‖ = 34
34 34 34 34

13

4	9	☐	16
15	☐	10	3
☐	7	11	2
1	12	8	13

14

1	14	☐	4
12	☐	6	9
8	11	10	5
☐	2	3	16

15 What do you notice about the three magic squares?

5

Let's investigate

13	8	12	1
2	11	7	14
3	10	6	15
16	5	9	4

Find sets of four numbers that add up to 34 in this magic square.

B

Attendance	
Ice skating	1159
Ice hockey	1084
Skiing	2057
Curling	768

Find the total number of tickets sold for these events.

1 ice skating and ice hockey

2 ice skating and curling

3 ice hockey and skiing

4 ice hockey and curling

5 skiing and curling

6 ice skating, ice hockey and skiing

7 37 more people went to watch ice hockey after the match had started. What is the total now?

Use the attendance numbers.

8 Round the numbers to the nearest 100.

9 Write the 3 most popular sports in order.

10 Use the rounded numbers to find the approximate total attending the 3 most popular sports.

11 If each player scores another 9 points, find the scores.

12 If each player scores another 99 points, find the scores.

Add these scores.

13 1256 and 387   **14** 1134 and 509

What do you notice?

15 399 + 401   **16** 999 + 1001

17 499 + 1401   **18** 799 + 1101

This is a magic triangle.
Every line of numbers
adds up to 20.

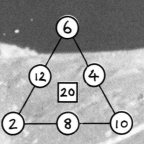

Complete these magic triangles.

19 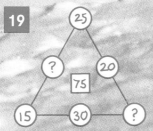   **20**

Let's investigate

Use the numbers 1 to 6 to make every
line of a magic triangle add up to 10.

Draw 2 more magic triangles.
One must add up to 100
and the other add up to 1000.

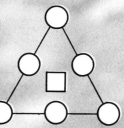

7

C

The first modern Olympic Games were held in 1896. They are held every 4 years. There were no games in 1916, 1940 and 1944 because of wars.

1 Write in order the years when the modern Olympic Games have been held starting with the first.

2 Write the years of the next 6 Olympic Games.

Gold	Silver	Bronze
180	226	216

These are the medals Great Britain has won in all the Olympic Games up to 1988.

3 How many medals have been won by Great Britain altogether?

Let's investigate

The years of the Olympic Games will all divide by 4.
Explain a way for finding numbers which will divide by 4.

Write some Olympic years and where the games were held.

Old pennies like these
were last used in 1971.
There were 240 pennies in £1.

A

1967   1900   1896

1877   1918   1946

1   Write the dates of the pennies in order.
    Start with the oldest.

2   What was the date 100 years before each coin?

3   What was the date 1000 years before each coin?

4   Look at the dates of these coins.
    What was the date 99 years before each one?

Half-crowns

1889   1921

Silver threepenny bits were
first issued in 1839.
They were replaced by
12-sided nickel-brass coins
which were last used in 1971.

1893   1938

Silver threepences

Silver sixpences

1957

1939

1904

1873

1919

1929

1895

1948

5 What is the date of the oldest coin?

6 What is the date of the newest one?

7 What is the difference between the dates?

What is the difference between each pair of dates?

8 1957 and 1929   **9** 1929 and 1904   **10** 1948 and 1919

11 1939 and 1895   **12** 1957 and 1948   **13** 1919 and 1895

Copy and finish these patterns of dates.

14 1954   1944   1934   ⬜ ⬜ ⬜

15 1940   1920   1900   ⬜ ⬜ ⬜

16 1938   1933   1928   ⬜ ⬜ ⬜

All British coins are made at the Royal Mint.
This used to be in London but moved to Wales in 1968.

Round each date to the nearest 100 years.

17 1782   **18** 1804   **19** 1892   **20** 1794   **21** 1879

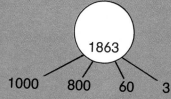

1863

1000   800   60   3

The date on the coin can be written as four numbers.

Write four numbers for each of these dates.

22

1912

1000   900   ■   ■   ■

23

1891

800   ■

24

1934

1000   ■   ■

Let's investigate

How many different dates can you make from these numbers?
Use four numbers each time.

1000   800   600   500   70   10
80   1   9   5

Put the dates in order. Start with the oldest.

11

B

People used to cut a penny in four to make 'fourthings' or farthings.
The first copper farthings were made in the reign of James I.

This is the last year farthings were minted.

1799
George III

1911
George V

These were the dates on farthings minted in the first year of the reigns of these kings and queens.

1 Write the dates of the coins in order. Start with the oldest.

What is the difference between the dates on the farthings of these kings or queens?

1838
Victoria

1953
Elizabeth II

2 Edward VII and William IV

3 George VI and Elizabeth II

4 William IV and George V

5 Edward VII and George VI

6 George V and George IV

1937
George VI

1831
William IV

7 In 1952 a 50 year old farthing was found. Which king's or queen's head would be shown on it?

8 If a 60 year old coin was found in 1971, which king's or queen's head would be shown on it?

1821
George IV

1902
Edward VII

9 What is the difference between the date on the oldest farthing on the page and the last year farthings were minted?

1886
half-crown

1938
half penny

1921
two shilling piece

1884
penny

1952
threepence

1893
sixpence

1946
farthing

1879
shilling

10 In which century was each coin minted?

11 Round the date on each coin to the nearest 10 years.

12 Round the date on each coin to the nearest 100 years.

1 9 ②①

The two ringed numbers in this date are worth 20 and 1.
Their difference is 19.

$$20 - 1 = 19$$

Do the same for these dates.

13 ①⑧8 4

14 1 ⑨⑤2

15 1 8 ⑨③

16 ①⑨4 6

17 1 ⑧⑦9

18 1 9 ③⑧

19 A coin found in 1957 was 93 years old.
What was the date on the coin? How old is it now?

Let's investigate

Use or draw some coins.
Write questions about their dates. Work out the answers.
Ask a friend to answer your questions.

C Find the missing numbers in these dates.

1
```
  1 9 5 4
- 1 8 ■ 7
  ─────────
    1 2 7
```

2
```
  1 9 7 ■
- 1 7 3 5
  ─────────
    2 3 5
```

3
```
  1 8 6 2
- 1 6 4 ■
  ─────────
    2 1 6
```

4
```
  1 9 ■ 6
- 1 6 2 8
  ─────────
    3 4 8
```

These dates make three number patterns.
Each pattern has five dates in it.
Write the dates in each pattern.

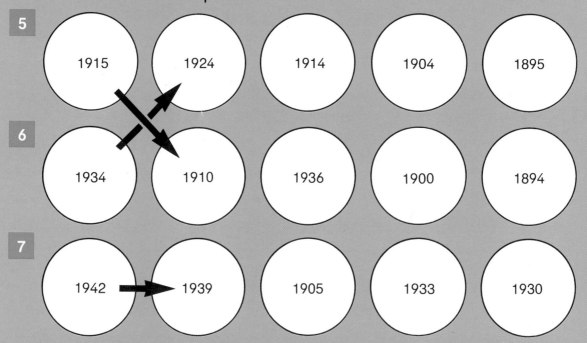

5 1915   1924   1914   1904   1895

6 1934   1910   1936   1900   1894

7 1942   1939   1905   1933   1930

Let's investigate

Make three patterns of dates.
Each pattern must have four dates in it.
Mix the dates up.
Ask a friend to find the three patterns.

Shape 1

This pattern of
crayons has two
lines of symmetry.

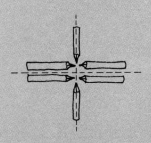

Use a mirror to
show them.

How many lines of symmetry can you find in
these patterns?

7 What are the shapes made with the cut sticks?

8 Use a mirror to find the lines of symmetry for each shape.

9 Draw round a template of each shape. Mark all the lines of symmetry.

10 Copy and complete this chart.

shape	△	□	⬠	⬡
number of lines of symmetry	3	4	5	6

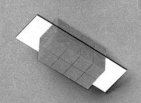

11 What pattern do you find?

12 What shape do the red cubes make? cuboid

13 The shape can be split so that one half reflects the other. The split is the plane of symmetry.

How many cubes are on each side of the mirror? 8

Let's investigate

Make some cuboids using small cubes.
Find different planes of symmetry.
Use a mirror to check.

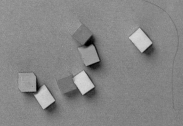

Quilling is a very old craft.
Strips of coloured paper are rolled
and glued to make shapes and patterns.
Quill work was often used to decorate
tops of workboxes and tea caddies.

How many lines of symmetry do these patterns have?

1

2

3

These models are made with polyspheres.
How many planes of symmetry does each one have?
Make models from plasticine and sticks to help you.

4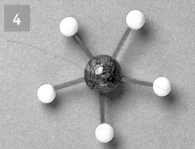

5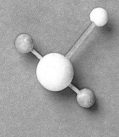

6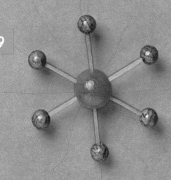

7

8

9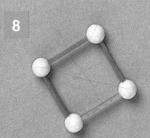

17

10 What shape are these?

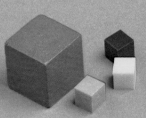

11 Use small cubes to make a bigger cube.
You must be able to split it to show
3 planes of symmetry.
Use a mirror to check.

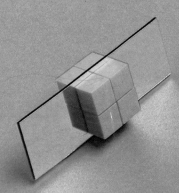

12 Describe any planes of symmetry
you cannot show by splitting the cube?

13 Make this shape
with cubes.

How many planes of
symmetry can you find?

Let's investigate

Use cubes to make shapes which you can
split on only one plane of symmetry.
Record your results. What do you notice
about the number of cubes you use?

C

1 Fold some circles in halves.
Cut out shapes so the circles look like
these when they are unfolded.

2 Fold some circles in quarters.
Cut out shapes to make the same
circle patterns when they are unfolded.

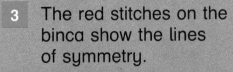

3 The red stitches on the
binca show the lines
of symmetry.

Use a mirror to help you
draw the whole patterns
on squared paper.

Let's investigate

Make or draw some cross stitch
patterns for your friend to complete.
Show one quarter of each pattern
for your friend to finish.
Use different coloured stitches.

Data 1

A   These shapes have been sorted using a Venn diagram.

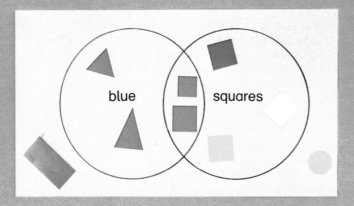

1   Why are two of the shapes not in the circles?

2   Why are two shapes in both circles?

3   The white shape is a ＿＿ but not ＿＿ .

4   The triangles are ＿＿ but not ＿＿ .

5   Copy this Venn diagram.
Sort these shapes into it.

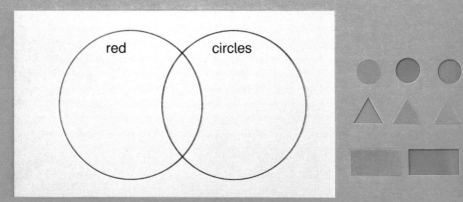

6   Copy and complete the
chart to explain what
you have done.

▲	Red and not a circle
●	Not red but a circle
■	Not red and not a circle

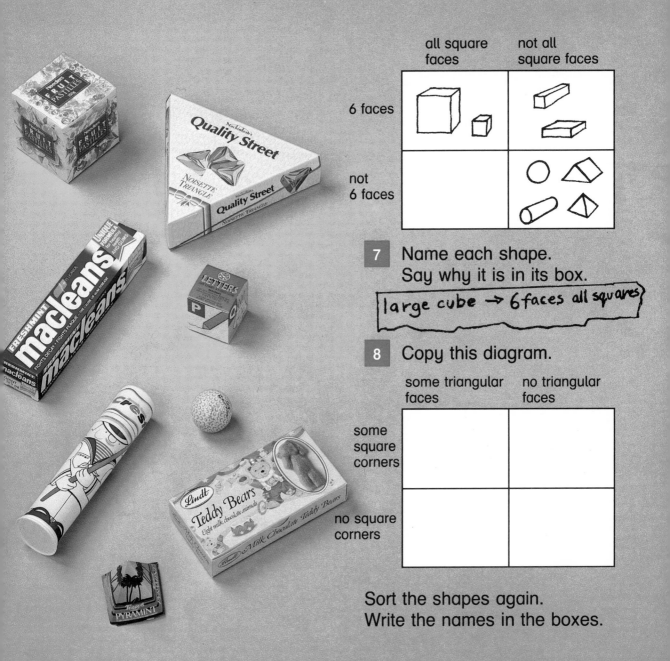

	all square faces	not all square faces
6 faces		
not 6 faces		

7 Name each shape.
Say why it is in its box.

large cube → 6 faces all squares

8 Copy this diagram.

	some triangular faces	no triangular faces
some square corners		
no square corners		

Sort the shapes again.
Write the names in the boxes.

Let's investigate

Draw the Venn diagram.
Label the circles.

Draw a set of 5 shapes so that
there are shapes in A, B and C.
Do it again in a different way.

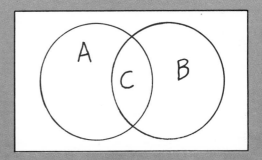

A   C   B

B

1 Decide which road each shape goes along.
Which box does each shape fit into?
Think of a way to record your answers.

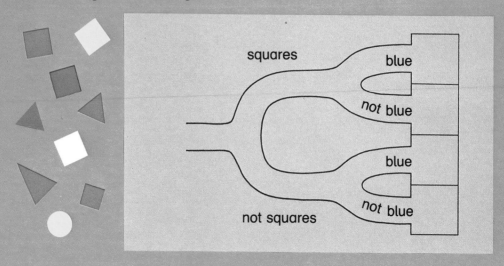

2 Do the same for these solid shapes.

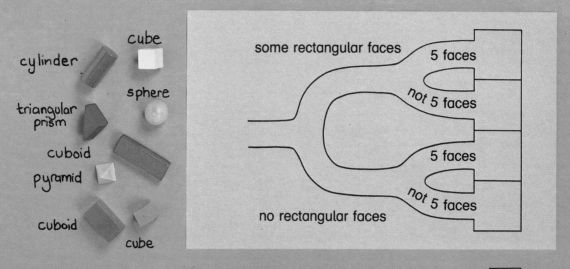

Use some shapes. Copy the road.
Write directions for the shapes
to get into the boxes.

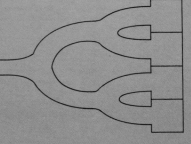

C

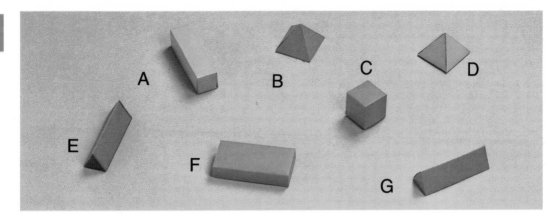

1 Copy this Venn diagram.
Sort the shapes into it.
Write the letters for each shape.

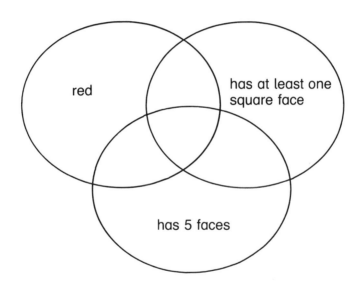

red

has at least one square face

has 5 faces

Let's investigate

Choose a set of shapes
that will fit into
this Venn diagram.

Sort the same shapes into
other Venn diagrams like
this but with different labels.

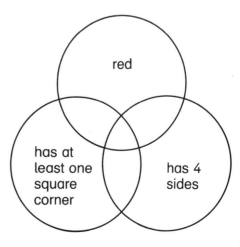

red

has at least one square corner

has 4 sides

23

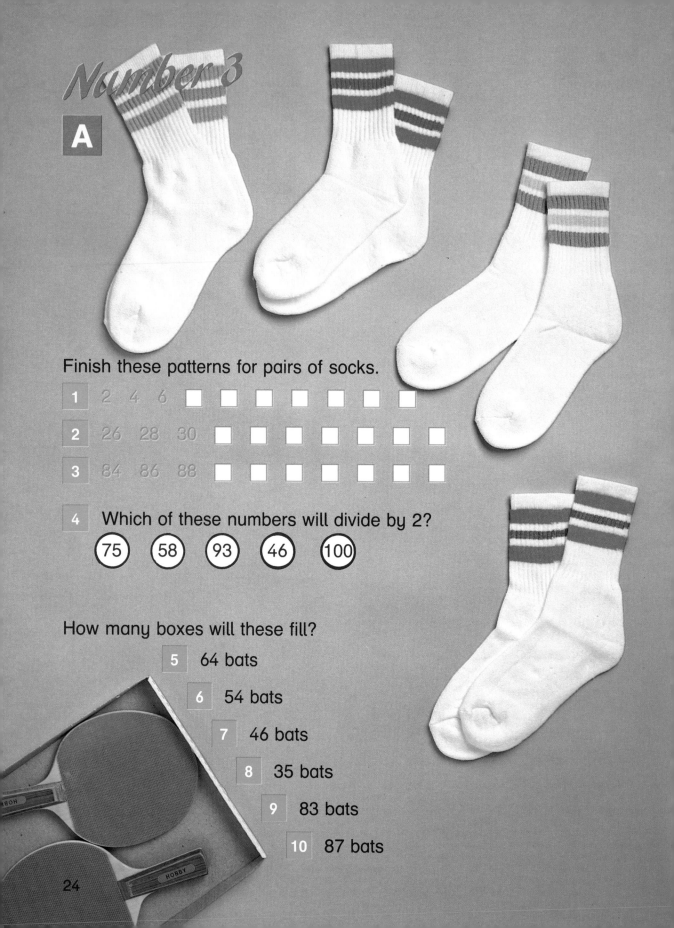

Number 3

A

Finish these patterns for pairs of socks.

1 2 4 6 □ □ □ □ □ □ □

2 26 28 30 □ □ □ □ □ □ □

3 84 86 88 □ □ □ □ □ □ □

4 Which of these numbers will divide by 2?

(75) (58) (93) (46) (100)

How many boxes will these fill?

5 64 bats

6 54 bats

7 46 bats

8 35 bats

9 83 bats

10 87 bats

24

Finish these patterns for bags of balls.

11 5 10 15 ☐ ☐ ☐ ☐ ☐ ☐ ☐

12 60 65 70 ☐ ☐ ☐ ☐ ☐ ☐ ☐

13 125 130 135 ☐ ☐ ☐ ☐ ☐ ☐ ☐

14 What do you notice about the pattern of 5?

How many bags of 5 will these balls fill?

15 45 balls   **16** 60 balls   **17** 85 balls

18 54 balls   **19** 83 balls   **20** 99 balls

How many triangles are needed
for these red snooker balls?

21 20 balls   **22** 50 balls

23 70 balls   **24** 100 balls

25 130 balls   **26** 180 balls

35 90 78 160 43 55

27 Which of these numbers will divide exactly by 5?

28 Which of these numbers will divide exactly by 10?

Let's investigate

Find the smallest number that will divide exactly by 2, 5 and 10.

Find some larger numbers that will.

What do you notice about the numbers you have discovered?

B

Finish these patterns for sets of darts.

1 3 6 9 ☐ ☐ ☐ ☐ ☐ ☐ ☐

2 36 39 42 ☐ ☐ ☐ ☐ ☐ ☐ ☐ ☐

3 72 75 78 ☐ ☐ ☐ ☐ ☐ ☐ ☐ ☐

How many sets will these darts make?

4 33 darts   **5** 51 darts   **6** 84 darts

7 40 darts   **8** 58 darts   **9** 77 darts

These numbers all divide exactly by 3. Add their digits.

10 30→   **11** 45→   **12** 24→

13 66→   **14** 72→   **15** 96→

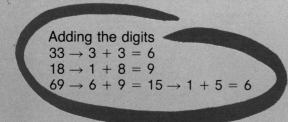

Adding the digits
33 → 3 + 3 = 6
18 → 1 + 8 = 9
69 → 6 + 9 = 15 → 1 + 5 = 6

16 What do you notice?

Find ten numbers between 100 and 200 which will divide exactly by 3.

Find ten numbers larger than 400 which will divide exactly by 3.

What do you notice about the sum of the digits of each number?

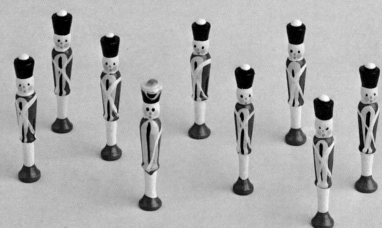

C

How many complete sets of
9 skittles can be made from these?

1	81 skittles	2	95 skittles
3	167 skittles	4	158 skittles
5	194 skittles	6	325 skittles

These numbers all divide exactly by 9. Add their digits.

7 45 **8** 99 **9** 198 **10** 288

11 What do you notice?

Complete the boxes so that the
number will divide exactly by 9.

$9\overline{)4\ \square\ \square}$

Find as many ways as possible.

Money 1

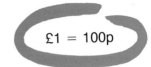

£1 = 100p

Draw the smallest number of coins
to pay for these books.

1 £1·50 → 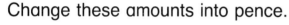 ☐   **2** £3·65

3 £1·95   **4** £2·99   **5** £2·50

Change these amounts into pence.

6 £1·50 = ☐ p   **7** £3·65   **8** £1·95

9 £2·99   **10** £2·50   **11** £4·95

12 What is the bill for the two cheapest books?

13 What is the bill for the two most expensive books?

How much do I pay for these books?

14 *China* and *Food*   **15** *Morning Break* and *Forts*

16	£	**17**	£	**18**	£
	2·50		1·99		4·25
+	3·75	+	4·00	+	3·75

19 How much has each child saved?

David

Jane

Ajit

Naomi

£2.25

The Witch's Brew and other poems
Wes Magee

20 Who can afford to buy *Blood and Guts?*

£1.99

ALLSORTS
আবরারের ছুটির দিনগুলো
Abrar's holiday
Audrey Fletcher

Illustrated by Beryl Sanders
Bengali version by Kalyan Adhar

21 Jane buys *The Witch's Brew.*
How much money has she left?

£5.70

ANDY
An Alaskan Tale
Susan Welsh-Smith   Rie Muñoz

22 David buys *Abrar's Holiday.*
How much money does he have left?

A PLACE FOR OWLS
True Animal Stories

23 Ajit buys *Andy.*
How much money does he have left?

24 Naomi buys *A Place for Owls.*
How much does she have left?

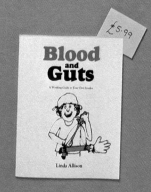

£5.99

Blood and Guts
A Working Guide to Your Own Insides
Linda Allison

Let's investigate

Tom spent £5 in the book sale.
He bought 3 books. He had 10p change.
Make up possible prices for the books he bought.

 B

The book tokens were prizes in a competition.
Sue's token was £6. Nadia had £7 and Tom won £9.

1 Look at the bookshelf below.
Choose 2 books for Sue.
Choose 3 books for Tom.
Show the bill and amount left for each child.

2 Whose books cost least?

3 Whose books cost most?

4 What is the difference between their bills?

£4·95 each   £3·65 each   £2·99 each   £2·50 each   £1·95 each

5 Copy and complete the price chart.

Number of books	1	2	3	4	5	6
Cost	£2·50	£5·00				

Make price charts for these two books.

6 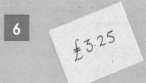 £3.25

7 £3.60

This is a till receipt.
It shows different things.

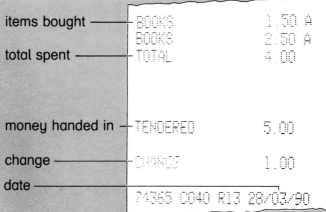

items bought ———— BOOKS   1.50 A
                  BOOKS   2.50 A
total spent ————— TOTAL   4.00

money handed in — TENDERED   5.00

change ————————— CHANGE   1.00
date ————————————
                  74365 C040 R13 28/03/90

Copy these receipts and fill in the missing amounts.

8
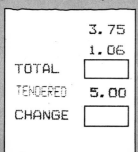
```
                3.75
                1.06
TOTAL      [      ]
TENDERED   5.00
CHANGE     [      ]
```

9
```
                1.99
                1.47
                0.74
TOTAL      [      ]
TENDERED   5.00
CHANGE     [   ]
```

10
```
                3.07
                0.85
           [      ]
TOTAL      8.99
TENDERED   10.00
CHANGE     [      ]
```

31

Round these prices to the nearest pound.

| 11 | 80p | 12 | £2·55 | 13 | £3·40 |
| 14 | £7·10 | 15 | £2·15 | 16 | £5·70 |

£2·67→£3

£1·34→£1

17 Round the prices of the books to the nearest pound and estimate the total cost.

Find the exact cost of the books. How close was your estimate?

£2·15

£3·90

Let's investigate

Find the correct price of 3 books you would like to buy. Which notes and coins would you choose to pay for all the books? Find different ways of paying. Do the same for 3 more books.

C

Let's investigate

You need a book catalogue and a calculator. If you had £50 to spend on the class library, what books would you choose?

Make a list and work out the total cost.

How much money is left?

Number 4

A thermometer measures the
temperature in degrees Celsius.
100 degrees Celsius or 100°C
is when water boils.
0°C is when water freezes.
10° below freezing is −10°C.

A

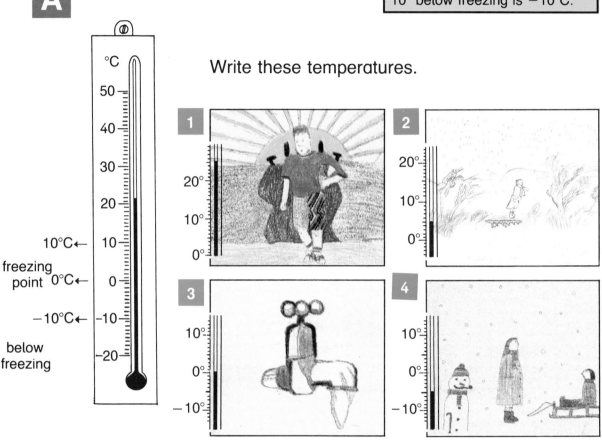

Write these temperatures.

5   What is the temperature in your classroom today?

6   Show these weather temperatures on a thermometer.

    3°C   −1°C   1°C   −3°C   2°C   −2°C

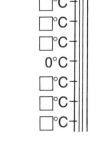

Let's investigate

Copy and finish this pattern −6 −3 0 ☐ ☐ ☐.

Write some patterns of your own. Start below zero.
Choose numbers between −12 and 20.

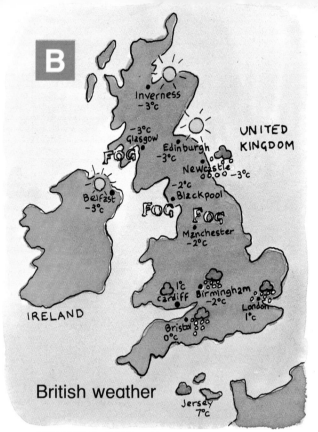

B British weather

Inverness −3°c
Glasgow −3°c
Edinburgh −3°c
UNITED KINGDOM
FOG
Newcastle −3°c
Belfast −3°c
Blackpool −2°c
FOG
FOG
Manchester −2°c
IRELAND
Cardiff 1°c
Birmingham −2°c
Bristol 0°c
London 1°c
Jersey 7°c

1. Which season do you think it is?

2. Which is the coldest place with snow?

3. Which is the warmest place with snow?

4. Which is the coldest foggy place?

5. Which are the coldest places with sunshine?

6. Which is the warmest place?

7. Which place has a temperature at freezing point?

Around the world December 12th					
Amsterdam	0°C	Cairo	21°C	Melbourne	30°C
Bahrain	23°C	Moscow	−3°C	Stockholm	−7°C
Chicago	−2°C	Dublin	−1°C	Hong Kong	21°C
Brussels	2°C	Helsinki	−15°C	Reykjavik	−9°C

8. Write the temperatures in order. Start with the hottest.

9. Show those temperatures which are below 0°C on a thermometer chart.

10. Why do you think Melbourne is so much hotter than Helsinki? Look at a world map for help.

Let's investigate

Which places in the world do you think will have temperatures even lower than those on the chart? Try to find their lowest temperatures and show them on a thermometer chart.

C

As you climb a mountain the temperature gradually falls. Even in hot countries some mountains have snow and ice on the top all year round.

1 The diagram below shows the height and temperatures of a 7000 m mountain.
Copy the diagram and write the temperatures.

The temperature drops about 6°C for every 1000 metres climbed.

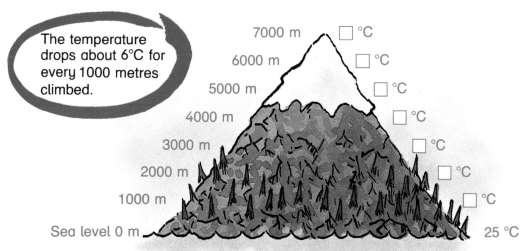

7000 m ☐ °C
6000 m ☐ °C
5000 m ☐ °C
4000 m ☐ °C
3000 m ☐ °C
2000 m ☐ °C
1000 m ☐ °C
Sea level 0 m   25 °C

2 After what height does the temperature go below freezing?

Let's investigate

Plan a diagram for different mountains less than 9000 m high.
The temperature at the summit of each one must be below zero.
It must not be hotter than 20°C at sea level.

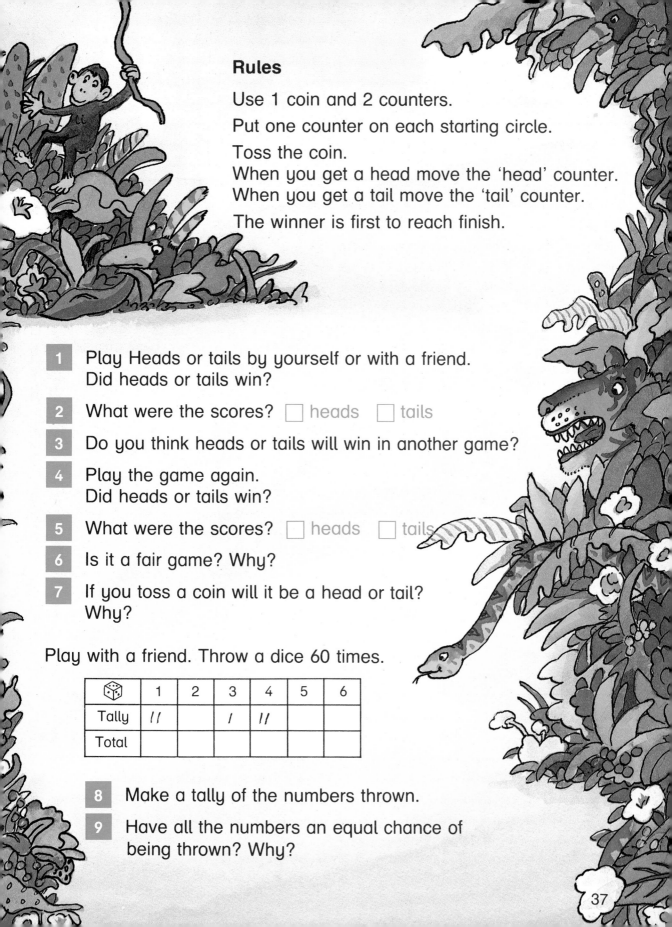

Rules

Use 1 coin and 2 counters.

Put one counter on each starting circle.

Toss the coin.
When you get a head move the 'head' counter.
When you get a tail move the 'tail' counter.

The winner is first to reach finish.

1. Play Heads or tails by yourself or with a friend.
   Did heads or tails win?

2. What were the scores? ☐ heads ☐ tails

3. Do you think heads or tails will win in another game?

4. Play the game again.
   Did heads or tails win?

5. What were the scores? ☐ heads ☐ tails

6. Is it a fair game? Why?

7. If you toss a coin will it be a head or tail?
   Why?

Play with a friend. Throw a dice 60 times.

🎲	1	2	3	4	5	6
Tally	//		/	//		
Total						

8. Make a tally of the numbers thrown.

9. Have all the numbers an equal chance of
   being thrown? Why?

Choose the best card for each of these sentences.

10 I shall laugh tomorrow.

11 I shall spend more than 10p today.

12 I shall be in bed at 6 p.m. today.

13 I shall be awake at 11 p.m. tonight.

14 I shall be dressed by 8 a.m. tomorrow.

15 I shall see my friends tomorrow.

16 I shall listen to the radio today.

Let's investigate

Write some sentences like these for each of the cards shown in the picture.

B

ODDS and EVENS

Rules

Use 1 dice and 2 counters.

Put one counter on each circle.

Throw the dice.
If it is even move the evens counter one forward. If it is odd move the odds counter one forward.

Keep playing until one counter passes the finish line.

1. Play Odds and evens by yourself or with a friend. Did odd or even win?

2. Do you think odd or even will win in another game?

3. Play the game again. Did odd or even win?

4. Have odd and even got an equal chance of winning? Why?

even
2 4
6

odd
1 3
5

Let's investigate

Use 1 dice and counters. Make up a game of your own. It must be fair. Explain your game.

finish

C SUM UP

Rules

Use two counters both marked 1 on one side
and 2 on the other, and some cubes.
Throw both counters and add the numbers.
Put a cube on the coloured circle for the score.
Keep throwing the counters and adding a cube.
The winning score is the one with the most cubes.

total
score
2

total
score
3

total
score
4

1   Play Sum up by yourself or with a friend for 3 minutes.
Which score won?

2   Write which score you think will win in another game.

3   Play the game again. Which score won?

4   Have all the scores an equal chance of winning? Why?

5   Is it possible to score 1? Why?

Let's investigate

total
score
☐ or ☐

total
score
☐

Use the two counters
marked 1 and 2 again
and some cubes.
Make up a game that
is fair using just
2 coloured circles
for the scores.
Make up an unfair game.
Explain why it is unfair.

Number 5

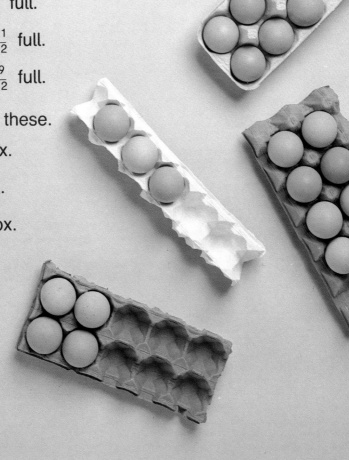

A

The full box has 6 eggs.
One pale egg is $\frac{1}{6}$ or 1 out of 6 of the eggs.

1 The pink box is $\frac{\square}{6}$ full.

2 The yellow box is $\frac{\square}{6}$ full.

3 Draw a box $\frac{2}{6}$ full of eggs.

4 Draw a box $\frac{5}{6}$ full of eggs.

5 The cream box holds 12 eggs altogether.
The box is $\frac{\square}{12}$ full.

6 Draw a box $\frac{11}{12}$ full.

7 Draw a box $\frac{9}{12}$ full.

Write fractions for these.

8 The white box.

9 The blue box.

10 The green box.

Draw pictures to show these.

11 $\frac{4}{5}$

12 $\frac{9}{10}$

13 $\frac{2}{5}$

41

14 Copy and finish each fraction chart.
Use the charts to answer the questions.

1 whole					
$\frac{1}{6}$	$\frac{1}{6}$				
$\frac{1}{12}$ $\frac{1}{12}$ $\frac{1}{12}$					

1 whole					
$\frac{1}{5}$	$\frac{1}{5}$				
$\frac{1}{10}$ $\frac{1}{10}$ $\frac{1}{10}$					

15 $\frac{2}{6}$ is the same as $\frac{\square}{12}$

16 $\frac{4}{6}$ is the same as $\frac{\square}{12}$

17 $\frac{5}{6}$ is the same as $\frac{\square}{12}$

18 $\frac{2}{12}$ is the same as $\frac{\square}{6}$

19 1 whole is the same as $\frac{\square}{5}$

20 $\frac{1}{5}$ is the same as $\frac{\square}{10}$

21 $\frac{3}{5}$ is the same as $\frac{\square}{10}$

22 $\frac{4}{10}$ is the same as $\frac{\square}{5}$

23 What is half of 6 eggs?
How did you do it?

24 What is $\frac{1}{6}$ of 6 eggs?
How did you do it?

25 What is $\frac{1}{5}$ of 10 eggs?
How did you do it?

Copy the eggs.

26  Ring $\frac{1}{10}$ of the eggs.

27

Ring $\frac{1}{2}$ of the eggs.

28

Ring $\frac{1}{6}$ of the eggs.

Let's investigate

Draw a box of eggs that you can write all these fractions for.

$\frac{1}{2}$ of ☐ = ☐ $\frac{1}{5}$ of ☐ = ☐ $\frac{1}{10}$ of ☐ = ☐

Find other numbers of eggs that you can write these same fractions for.

B   What fractions of eggs are in egg-cups?

1    **2**

3    **4**

5   What is $\frac{1}{5}$ of 10 eggs?
What is $\frac{2}{5}$ of 10 eggs?
How did you do it?

6   What is $\frac{4}{6}$ of 6 eggs?
How did you do it?

7   Find $\frac{3}{10}$ of 20 eggs.
How did you do it?

8   Find $\frac{8}{12}$ of 12 eggs.

9   Find $\frac{5}{6}$ of 12 eggs.

This is $\frac{4}{6}$ of a strip.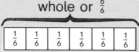

This is the whole strip.

Use squared paper.
Draw and cut out the whole strips for these.
Stick them in your book and label them.

10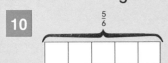

11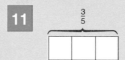

12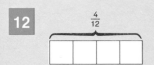

13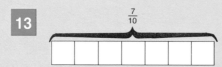

14 Copy this on to large squared paper.

$\frac{1}{3}$	$\frac{2}{3}$	$\frac{4}{6}$	$\frac{5}{10}$	$\frac{4}{8}$	$\frac{6}{12}$
$\frac{2}{6}$	$\frac{3}{4}$	$\frac{1}{3}$	$\frac{4}{12}$	$\frac{2}{6}$	$\frac{2}{4}$
$\frac{4}{12}$	$\frac{6}{8}$	$\frac{9}{12}$	$\frac{1}{5}$	$\frac{2}{10}$	$\frac{3}{6}$
$\frac{1}{6}$	$\frac{2}{12}$	$\frac{1}{4}$	$\frac{2}{8}$	$\frac{3}{12}$	$\frac{1}{2}$

Follow the path.

Colour every fraction equal to $\frac{1}{2}$ red.

Colour every fraction equal to $\frac{1}{3}$ blue.

Colour every fraction equal to $\frac{3}{4}$ yellow.

Colour every fraction equal to $\frac{2}{3}$ green.

Colour every fraction equal to $\frac{1}{6}$ white.

Colour every fraction equal to $\frac{1}{5}$ brown.

Colour every fraction equal to $\frac{1}{4}$ grey.

Use these charts to help you.

1							
$\frac{1}{2}$				$\frac{1}{2}$			
$\frac{1}{4}$		$\frac{1}{4}$		$\frac{1}{4}$		$\frac{1}{4}$	
$\frac{1}{8}$	$\frac{1}{8}$	$\frac{1}{8}$	$\frac{1}{8}$	$\frac{1}{8}$	$\frac{1}{8}$	$\frac{1}{8}$	$\frac{1}{8}$

1									
$\frac{1}{5}$		$\frac{1}{5}$		$\frac{1}{5}$		$\frac{1}{5}$		$\frac{1}{5}$	
$\frac{1}{10}$	$\frac{1}{10}$	$\frac{1}{10}$	$\frac{1}{10}$	$\frac{1}{10}$	$\frac{1}{10}$	$\frac{1}{10}$	$\frac{1}{10}$	$\frac{1}{10}$	$\frac{1}{10}$

1											
$\frac{1}{3}$				$\frac{1}{3}$				$\frac{1}{3}$			
$\frac{1}{6}$		$\frac{1}{6}$		$\frac{1}{6}$		$\frac{1}{6}$		$\frac{1}{6}$		$\frac{1}{6}$	
$\frac{1}{12}$	$\frac{1}{12}$	$\frac{1}{12}$	$\frac{1}{12}$	$\frac{1}{12}$	$\frac{1}{12}$	$\frac{1}{12}$	$\frac{1}{12}$	$\frac{1}{12}$	$\frac{1}{12}$	$\frac{1}{12}$	$\frac{1}{12}$

Let's investigate

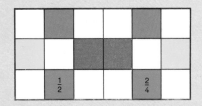

Copy and colour the pattern.
The pattern has symmetry.

Write equivalent fractions on squares of the same colour.
Don't use the same fraction twice.
Make another fraction pattern of your own that
also has symmetry.

C Write a fraction for each mark along
the line to fit the pattern.

1

$0 \quad \frac{1}{10} \quad \frac{2}{10} \quad \frac{3}{10} \quad \frac{\square}{\square} \quad \frac{1}{2} \quad \frac{\square}{\square} \quad \frac{\square}{\square} \quad \frac{\square}{\square} \quad \frac{\square}{\square} \quad 1$

2
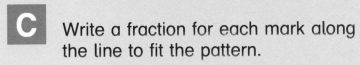
$0 \qquad \frac{\square}{5} \qquad \frac{\square}{\square} \qquad \frac{\square}{\square} \qquad \frac{\square}{5} \qquad 1$

3

$0 \qquad \frac{\square}{\square} \qquad \frac{\square}{\square} \qquad \frac{1}{2} \qquad \frac{\square}{\square} \qquad \frac{\square}{\square} \qquad 1$

4

$0 \qquad\qquad\qquad\qquad\qquad\qquad\qquad\qquad 1$

Let's investigate

An ostrich nest has 30 eggs in it.
What is $\frac{4}{5}$ of 30?
What is $\frac{8}{10}$ of 30?
What do you notice?

Find some more numbers and
fractions that give the same answers.

More than one ostrich will
lay eggs in the same nest.

45

A

INPUT

4
in a box

OUTPUT

This machine packs
pots into boxes.

Complete the charts
for this machine.

1

Input pots	8	24	32	40
Output boxes	2	☐	☐	☐

2

Input pots	12	☐	☐	☐
Output boxes	3	7	10	25

INPUT

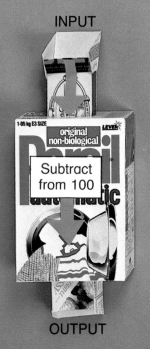

Subtract
from 100

OUTPUT

Put these numbers into this machine.
Find the outputs.

3 80   **4** 50

5 35   **6** 40

7 67   **8** 91

These are outputs from the machine.
Find the inputs.

9 25   **10** 40

11 5   **12** 15

13 35   **14** 55

15 Copy and finish the table for this machine.

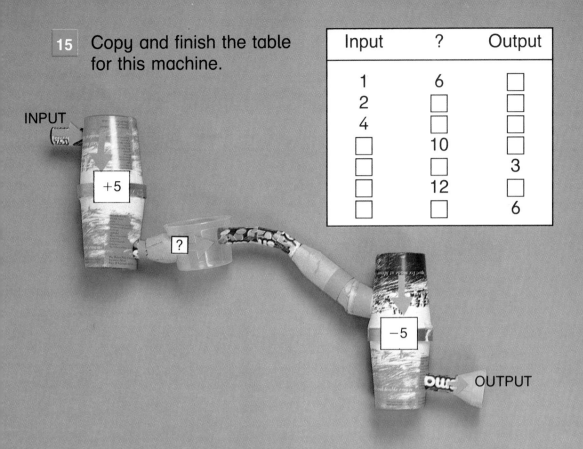

Input	?	Output
1	6	☐
2	☐	☐
4	☐	☐
☐	10	☐
☐	☐	3
☐	12	☐
☐	☐	6

INPUT

+5

?

−5

OUTPUT

Put these numbers into this machine. Find the output.

INPUT

Round numbers to the nearest 10

OUTPUT

16	67	**17**	93	**18**	81
19	89	**20**	98	**21**	31
22	124	**23**	208	**24**	497

Let's investigate

Design different machines of your own. Label them.
Make a table showing the input and output for each machine.

B Calculators are also number machines.

Find the missing numbers.
Use a calculator to check each time.

	Enter	Press	Shows
	4	⊞ ② ⊞ ① ⊟	7
1	3	⊞ ⑤ ⊟ ② ⊟	☐
2	5	⊠ ② ⊞ ① ⊟	☐
3	8	⊠ ③ ⊞ ☐ ⊟	27
4	12	÷ ② ⊞ ① ⊟	☐
5	24	÷ ⑧ ⊠ ☐ ⊟	6
6	30	÷ ⑤ ⊠ ☐ ⊟	18

This machine doubles both numbers of a fraction.

INPUT $\frac{1}{2}$ ... double double ... OUTPUT $\frac{2}{4}$

If these fractions go in, what comes out?

7 $\frac{1}{2} = \frac{\square}{\square}$ **8** $\frac{2}{4} = \frac{\square}{\square}$ **9** $\frac{4}{8} = \frac{\square}{\square}$

10 Copy this fraction chart and finish it.

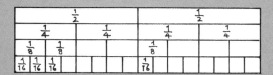

11 Write the fractions.

$\frac{1}{2} \rightarrow \frac{\square}{4} \rightarrow \frac{\square}{8} \rightarrow \frac{\square}{16}$

12 Colour them on the fraction chart.
What do you notice?

Use the fraction doubling machine again.
Write these fractions.

13 $\frac{1}{3} \rightarrow \frac{\square}{6} \rightarrow \frac{\square}{12}$ **14** $\frac{2}{3} \rightarrow \frac{\square}{6} \rightarrow \frac{\square}{12}$

15 Put these numbers in the double and halve machine. Complete the table.

INPUT

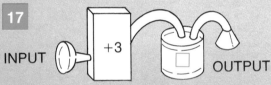
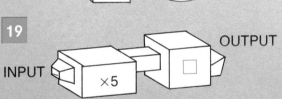

OUTPUT

Input	Double	Halve	Output
5	10	☐	☐
7	☐	☐	☐
10	☐	☐	☐
24	☐	☐	☐
35	☐	☐	☐

16 What happens to a number when you double and then halve it?

Complete these machines. Each number going in must be the same as the number coming out.

17

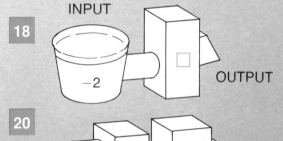

INPUT $+3$ OUTPUT

18

INPUT

-2 ☐ OUTPUT

19

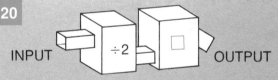

INPUT $\times 5$ OUTPUT

20

INPUT $\div 2$ ☐ OUTPUT

21 Copy and finish the table for this machine.

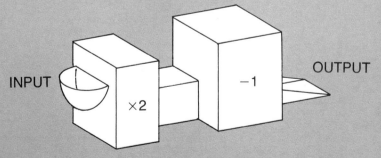

INPUT $\times 2$ -1 OUTPUT

Input	Output
3	5
10	☐
21	☐
☐	13
☐	17

22 Write what you notice about all the numbers coming out of this machine.

23 Find a number that is the same going in and coming out.

Let's investigate

Complete this machine
so that only even numbers
come out.
Find different ways to do it.

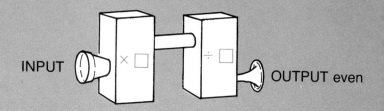

INPUT   ×☐   ÷☐   OUTPUT even

C   This machine does 3 things.

INPUT   × 2   × 2   × 5   OUTPUT

Complete these machines so that they do the same job.

1

INPUT   ×☐   ×☐   OUTPUT

2

INPUT   ×☐   OUTPUT

This machine
does 4 things.

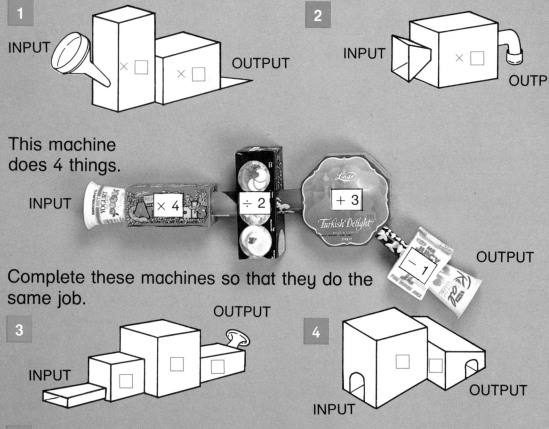

INPUT   × 4   ÷ 2   + 3   − 1   OUTPUT

Complete these machines so that they do the
same job.

3

OUTPUT

INPUT   ☐   ☐   ☐

4

INPUT   ☐   ☐   OUTPUT

5   If 8 came out of the machine, what was put in?

Write what these machines do.

6

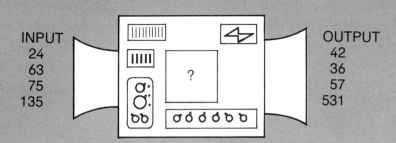

INPUT
24
63
75
135

OUTPUT
42
36
57
531

7

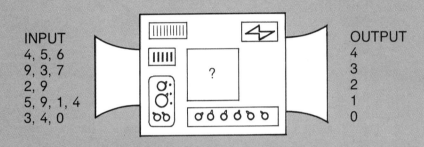

INPUT
4, 5, 6
9, 3, 7
2, 9
5, 9, 1, 4
3, 4, 0

OUTPUT
4
3
2
1
0

8

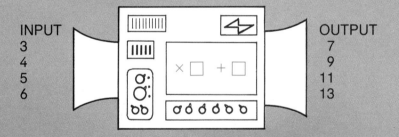

INPUT
3
4
5
6

×☐ + ☐

OUTPUT
7
9
11
13

Let's investigate

Make an instruction for this machine and show what numbers come out. Find other ways to do it. Explain them.

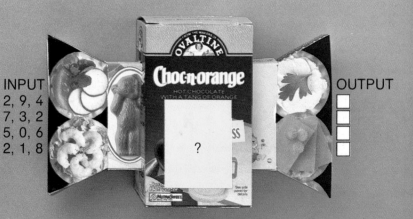

INPUT
2, 9, 4
7, 3, 2
5, 0, 6
2, 1, 8

OUTPUT

51

Length 1

Circle shapes are used in almost all the machines that we have today.

A

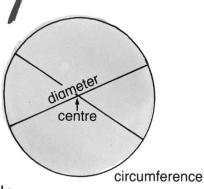

diameter
centre
circumference

Use a paper circle.

1 Fold it in half in different ways.

2 What is the line of each fold called?

3 Where do all the lines cross?

4 What is the outside edge of the circle called?

This model car takes tokens.
How long would the slots be
for these tokens?

5 10 LAPS

6 1 LAP

7 5 LAPS

Work with some friends.

8 Measure the circumference of a trundle wheel to the nearest centimetre.

9 The trundle wheel makes 10 turns. How many metres is this?

10 Copy the table.

Circle	Radius	Diameter
red	1cm	2cm
yellow		
green		
blue		

11 Measure the radius and diameter of these circles. Complete the table.

Radius is the distance from the centre to the circumference.

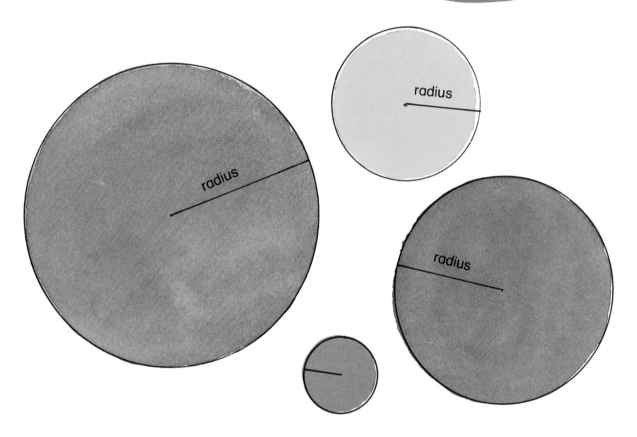

12 What do you notice about the radius and diameter of these circles?

Let's investigate

Use paper circles of different sizes.
Measure the radius and diameter of each.
What do you notice?

B

1   Copy the table.

Cylinders	Circumference		Diameter	
	Estimate	Measure	Estimate	Measure

2   Choose 4 cylinders.
Estimate the circumference of each in centimetres.
Write your estimates in the table.

3   Measure the circumferences.
Write the measurements in the table.

4   Estimate the diameter of each cylinder.
Write your estimates in the table.

5   Find a way to measure the diameters.
Complete the table.

2 m

The circumference of this bicycle wheel is 2 m.

How many turns has the wheel made when the bicycle has travelled these distances?

6 30 m   **7** 64 m   **8** 54 m

These rings have a circumference of 6 cm. How much wire is needed to make these?

9 12 rings   **10** 15 rings   **11** 20 rings

How many rings could these lengths of wire make?

12 180 cm   **13** 240 cm   **14** 3 metres

Let's investigate

Use card circles.

Find pairs where one diameter is half the other.
Measure their circumferences to the nearest centimetre.
What do you find?

Try it with other pairs of circles.

C This bicycle is called a penny-farthing. Find out why.

The circumference of the large wheel is 6 metres.
How far will the penny-farthing have travelled after the large wheel has turned these numbers of times?

1 1 turn   **2** 6 turns   **3** 20 turns

The circumference of the small wheel is 2 metres.
How many times will the large wheel have turned when the small wheel has turned these number of times?

4 6 times   **5** 15 times   **6** 36 times

7 Which tyre do you think will wear out first on the penny-farthing bicycle? Give your reason. How much faster will it wear out? Why?

8 Do tyres on cars with small wheels wear out slower or faster than on cars with larger wheels? Why?

Let's investigate

Use some lids or wheels.
Measure the circumference and diameter of each.
About how many times longer is the circumference than the diameter each time?
What do you notice?

Weight 1

× 5 g means each mark on the scale shows 5 g

A Write the weights shown on the scales.

1

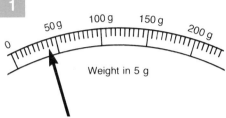

50 g   100 g   150 g   200 g
0

Weight in 5 g

The letter weighs ☐ g.

2

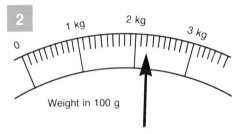

1 kg   2 kg   3 kg
0

Weight in 100 g

The parcel weighs ☐ kg ☐ g.

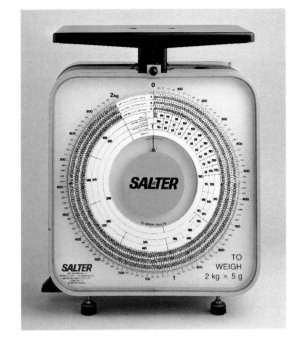

SALTER

SALTER   TO WEIGH 2 kg × 5 g

What weights do these spring balances show?

3

1 kg × 10 g
— 0
— 50
— 100
— 150
— 200
— 250
— 300
— 350
— 400
— 450
— 500
— 550

☐ g

4

1 kg × 10 g
— 0
— 50
— 100
— 150
— 200
— 250
— 300
— 350
— 400
— 450
— 500
— 55

☐ g

5

10 kg × 100 g
— 0
— 500
— 1 kg
— 500
— 2 kg
— 500
— 3 kg
— 500
— 4 kg
— 500
— 5 kg
— 500

☐ kg ☐ g

6

10 kg × 100 g
— 0
— 500
— 1 kg
— 500
— 2 kg
— 500
— 3 kg
— 500
— 4 kg
— 500
— 5 kg
— 500

☐ kg ☐ g

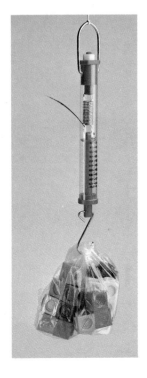

57

7 Work with some friends.

Use a spring balance.

Choose six things that you can weigh on it.
Weigh them and write their weights.

The _____ weighs ☐.

Make a stretch balance.

8 Fix a hook into a board.
Pin a piece of paper under the hook.
Hang a rubber band on the hook.
Put an S hook on the band.
Mark where the bottom of the
rubber band comes to.

9 Hang a key on the hook.
Mark where the bottom of the
band is now.

10 Hang a pair of scissors on the hook.
Mark where the band stretches to now.

11 Which stretches the band more, the
key or the scissors?
Why is this?

Let's investigate

Use your stretch balance.
Find out what happens if you use rubber bands
of different thicknesses.
Explain what you discover.

58

B What are the readings on these scales?
Write each weight in kg, then again in g.

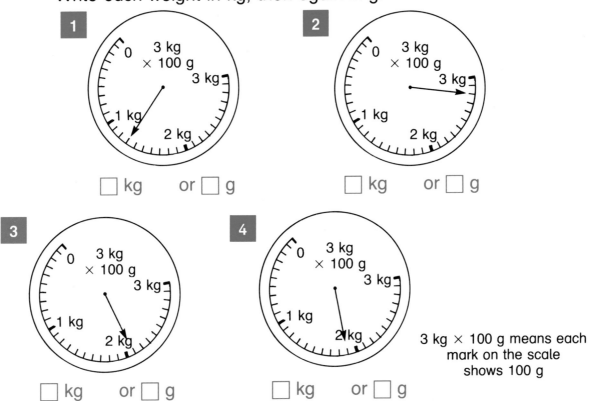

1
☐ kg   or ☐ g

2
☐ kg   or ☐ g

3
☐ kg   or ☐ g

4
☐ kg   or ☐ g

3 kg × 100 g means each
mark on the scale
shows 100 g

5 Choose four things that each weigh less than $2\frac{1}{2}$ kg.
Weigh them and show each weight on a bar chart.

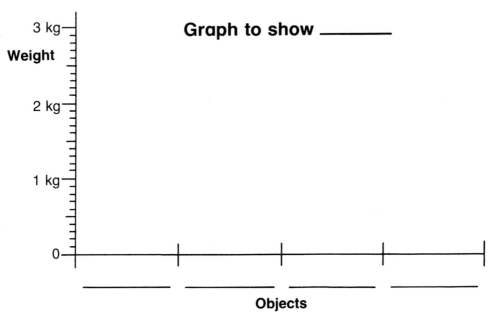

Graph to show _____

Weight

3 kg

2 kg

1 kg

0

Objects

59

6   Choose an object that you can weigh on a spring balance.
Write its weight in kg.

7   Find another object that you think will weigh
about the same as the first one.
Weigh it and write its weight.

8   What was the difference in the two weights?

9   Try this again with other pairs of objects.
Record your results each time.
Do you find any pairs that weigh the same?

Let's investigate

Make a see-saw balance with a ruler,
a cardboard tube and some Blu-Tak.

Hint: Use Blu-Tak to
fix the tube to your
desk first.

Put a 50 g weight on each end and make them balance.

Find a way to make a 50 g weight balance a 100 g weight.

Do the same with 50 g and 150 g.

Try it with other weights.

Draw your results.
Explain what you did and what you discovered.

C

1 Choose 5 different newspapers.
Weigh each one.

Show their weights in a table.

Newspaper					
Weight					

2 Which is the heaviest newspaper?

3 Which is the lightest?

4 Give some reasons why some newspapers are heavier than others.

Let's investigate

Find out how many papers and magazines are delivered to your house, or the house of a friend, each week.

Estimate their total weight.

Volume and capacity

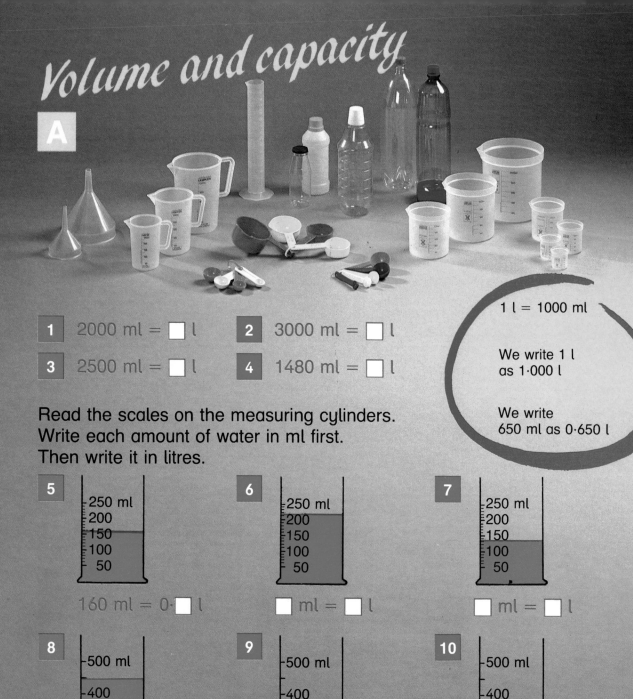

1   2000 ml = ☐ l   2   3000 ml = ☐ l

3   2500 ml = ☐ l   4   1480 ml = ☐ l

1 l = 1000 ml

We write 1 l
as 1·000 l

We write
650 ml as 0·650 l

Read the scales on the measuring cylinders.
Write each amount of water in ml first.
Then write it in litres.

5
- 250 ml
- 200
- 150
- 100
- 50

160 ml = 0·☐ l

6
- 250 ml
- 200
- 150
- 100
- 50

☐ ml = ☐ l

7
- 250 ml
- 200
- 150
- 100
- 50

☐ ml = ☐ l

8
- 500 ml
- 400
- 300
- 200
- 100

☐ ml = ☐ l

9
- 500 ml
- 400
- 300
- 200
- 100

☐ ml = ☐ l

10
- 500 ml
- 400
- 300
- 200
- 100

☐ ml = ☐ l

11 Find two containers that each hold less than 1 litre.
Measure their capacities.
Write each answer in ml and then in litres.

12 Work with a friend.

Pour 100 ml of water into a measuring cylinder.
Put in a block of 30 centimetre cubes.
They must be covered by the water.
You may have to hold them under with a straw.

The water level is now at ☐ ml.

13 Start again with 100 ml of water in the cylinder.
Put in a block of 20 centimetre cubes.

The water level is now at ☐ ml.

14 Do the same with 10 centimetre cubes.
The water level is at ☐ ml.

15 What do you think would happen if you
put just 1 centimetre cube in?

I think that 1 centimetre
cube will make the
water level rise ☐ ml.

Let's investigate

Make a ball of plasticine.
Find out how much space
it takes up in the water.

Find out how much space
some other things take up.
Record your findings.

B

Half fill a measuring jug with water.

1 The reading is ☐ ml.

Now put a stone in.

2 The new reading is ☐ ml.

3 The stone takes up the same
space as ☐ centimetre cubes.

4 The volume of the stone
is ☐ cubic centimetres.

5 Estimate the volumes of another two stones.
Use the measuring jug to check your estimates.
Write your results.
  Estimated volume is ☐ cubic centimetres.
  Measured volume is ☐ cubic centimetres.

6 Use different sized measuring jugs to find
the volume of the same stone.
What do you find?

Let's investigate

Half fill a measuring jug with water.

Drop a ball of plasticine into the water.
Check the water level.

Change the shape of the plasticine
and put it back in the water.
Check the water level now.

Do the same again.
What do you discover each time?

C

1 Weigh a piece of dry cloth.
Soak it in water.
Squeeze it gently until it just stops dripping.
Weigh it again.
How many grams of water was it holding?

Let's investigate

Find the weight of 1 ml of water.
Explain how you did it.
How many ml of water do you think
the wet cloth was holding?

Time 1

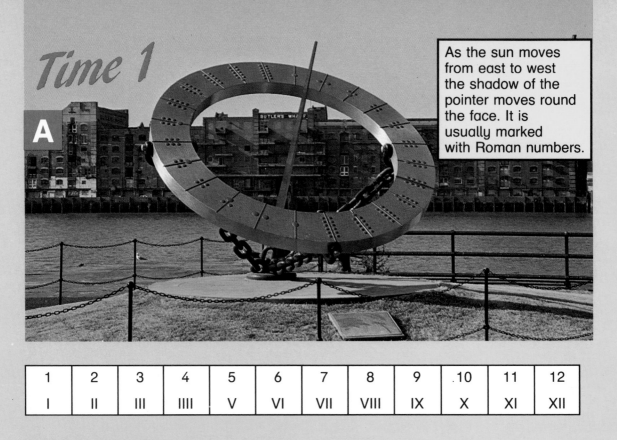

As the sun moves from east to west the shadow of the pointer moves round the face. It is usually marked with Roman numbers.

1	2	3	4	5	6	7	8	9	.10	11	12
I	II	III	IIII	V	VI	VII	VIII	IX	X	XI	XII

What are the times shown by these sundials?
The shadow points to the time.

1

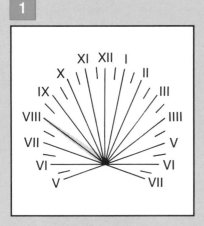

2

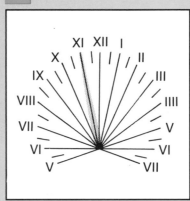

3

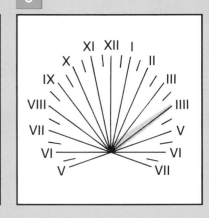

Hilaire Belloc wrote:

> I am a sundial and I make a botch
> of what is done far better by a watch.

4 Why are sundials not good for telling the time?

The lantern clock was one of the first indoor clocks. They were driven by heavy weights hung from cords in the centre. They were not very accurate.

Remember

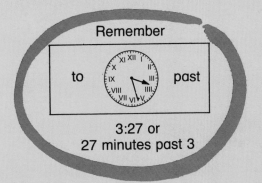

to   past

3:27 or
27 minutes past 3

Write the times shown on these clocks in two ways.

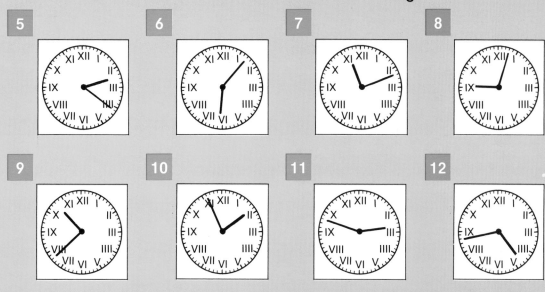

| 5 | 6 | 7 | 8 |
| 9 | 10 | 11 | 12 |

Let's investigate

The world record for running 100 m is just less than 10 seconds.
How far can you run in 10 seconds?
What else can you do in 10 seconds?

There are 60 seconds in 1 minute.

B

What are these Roman numbers?

1 VII   **2** IX   **3** V   **4** VI

5 Draw your own clock.
Put on Roman numbers.

What are these times?
Write your answers as digital times.

6   **7**   **8**

9   **10**   **11**

Show these times on clock faces with
Roman numbers.

12   **13** 3:28   **14** 10:51

15 8:03   **16** 12:26   **17** 9:17

A grandfather
clock is worked by
a swinging
pendulum.

This watch can be a stopwatch too. It can time in seconds. Races are timed in seconds.

Points chart	
Time (seconds)	Points
7	75
8	55
9	35
10	20
11	10
12	5
13	2
14	1

Find the total points for the four teams.

18

Red team	Time	Points
Abida	10 secs	20
Bob	9 secs	
Total points		

19

Yellow team	Time	Points
Zoe	8 secs	
David	14 secs	
Total points		

20

Blue team	Time	Points
Susan	12 secs	
Talika	9 secs	
Total points		

21

Green team	Time	Points
Emma	11 secs	
Ranjit	10 secs	
Total points		

22 Which team has most points?

Let's investigate

Choose an activity and make up your own points chart. Time 2 children doing it and work out their scores.

C

Swinging pendulums
make clocks accurate.
Most grandfather clocks
have pendulums that
swing once every second.
Cuckoo clock pendulums
swing faster.

1 Make a pendulum with plasticine and string.

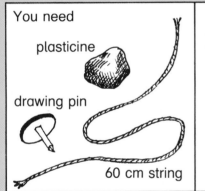

You need

plasticine

drawing pin

60 cm string

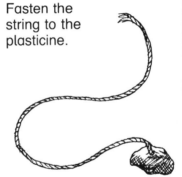

Fasten the
string to the
plasticine.

Fix the string so
that the pendulum
can swing.

2 How long does it take
for 10 complete swings?

The plasticine
must come back
to make one
complete swing.

3 How many complete swings are made in one minute?

Let's investigate

Try to make the pendulum swing once in 2 seconds.
Find out what changes the time of the swing.

A

 CAMBRIDGE PRIMARY MATHEMATICS

Multiplication squares

X	1	2	3	4	5	6	7	8	9	10
1	1	2	3	4	5	6	7	8	9	10
2	2	4	6	8	10	12	14	16	18	20
3	3	6	9	12	15	18	21	24	27	30
4	4	8	12	16	20	24	28	32	36	40
5	5	10	15	20	25	30	35	40	45	50
6	6	12	18	24	30	36	42	48	54	60
7	7	14	21	28	35	42	49	56	63	70
8	8	16	24	32	40	48	56	64	72	80
9	9	18	27	36	45	54	63	72	81	90
10	10	20	30	40	50	60	70	80	90	100

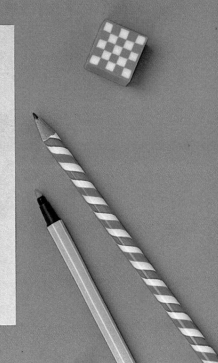

Colour in these answers on a multiplication square.

1
$2 \times 4 = \square$
$4 \times 2 = \square$

2
$8 \times 9 = \square$
$9 \times 8 = \square$

3
$3 \times 4 = \square$
$4 \times 3 = \square$

4
$6 \times 4 = \square$
$4 \times 6 = \square$

5
$4 \times 5 = \square$
$5 \times 4 = \square$

6
$8 \times 7 = \square$
$7 \times 8 = \square$

7
$7 \times 5 = \square$
$5 \times 7 = \square$

8
$10 \times 8 = \square$
$8 \times 10 = \square$

9 Draw in the diagonal line on your multiplication square.
Write about the pattern you see.

10
$21 \times 3 = 63$
$63 \div 3 = \square$

11
$23 \times 2 = \square$
$46 \div 2 = \square$

12
$16 \times 3 = \square$
$48 \div 3 = \square$

13
$32 \times 3 = \square$
$96 \div 3 = \square$

14
$27 \times 2 = \square$
$54 \div 2 = \square$

15
$21 \times 4 = \square$
$84 \div 4 = \square$

16 What do you notice about the numbers each time?

17
$221 \times 4 = \square$
$884 \div 4 = \square$

18
$117 \times 2 = \square$
$234 \div 2 = \square$

19
$132 \times 3 = \square$
$396 \div 3 = \square$

You can show the table
of 9 on your fingers.

$3 \times 9 = 27$

$(3) \times 9 = (20) + (7)$

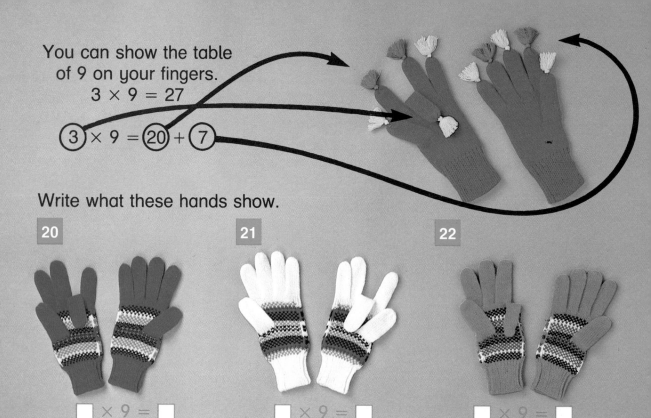

Write what these hands show.

20

$\square \times 9 = \square$

21

$\square \times 9 = \square$

22

$\square \times 9 = \square$

We can find 8 × 11 by doubling.	To find 5 × 11 we add these.
double double	
$1 \times 11 = 11$ $2 \times 11 = 22$ $4 \times 11 = 44$ $8 \times 11 = 88$	$1 \times 11 = 11$ $\underline{4 \times 11 = 44}$ add $5 \times 11 = 55$

Use the doubling table of 11. Do these by adding.

23 $3 \times 11 = \square$ **24** $6 \times 11 = \square$ **25** $9 \times 11 = \square$

Let's investigate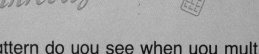

What pattern do you see when you multiply by 10?
Use a calculator to find a pattern when you multiply by 100.
Explain why you get the pattern.

B

These are Napier's rods.

1 Copy them to make your own set.

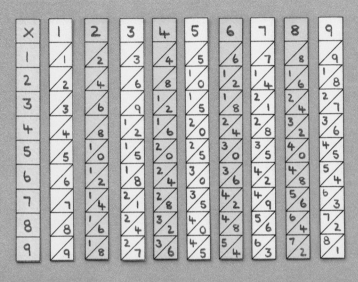

We can use them to multiply.

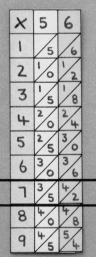

To find 56 × 7 we put out the 5 and 6 rods and also the × rod.
Look across from the 7 on the × rod.

Add the numbers diagonally.
56 × 7 = 392

Use Napier's rods to do these.

2 35 × 5 = ☐   **3** 47 × 6 = ☐   **4** 21 × 7 = ☐   **5** 93 × 3 = ☐

6 63 × 4 = ☐   **7** 72 × 4 = ☐   **8** 59 × 6 = ☐   **9** 85 × 9 = ☐

Check your answers with a calculator.

To find 138 × 3 we put
out the 1, 3 and 8 rods
and the × rod.
Look across from the 3.

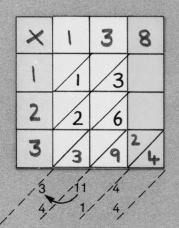

Add the numbers.
138 × 3 = 414

Use Napier's rods to do these.

10 154 × 5 = ☐   **11** 327 × 3 = ☐   **12** 369 × 2 = ☐

13 498 × 6 = ☐   **14** 386 × 7 = ☐   **15** 419 × 6 = ☐

Check your answers in another way.

Let's investigate

Here is another way of multiplying 254 × 3.

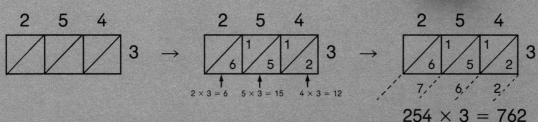

254 × 3 = 762

Make up some multiplications of your own like this.
Check them on a calculator.

We can multiply in many different ways.
Here are two ways.

```
    1  2  1              1  2  1
  ×        8           ×        8
    9  6  8                    8 ——   1 × 8
                        1  6  0 —— 2 0 × 8
                        8  0  0 ——1 0 0 × 8
                        9  6  8 ——1 2 1 × 8
```

Use both ways to do these.

| **1** | 136 × 8 = ☐ | **2** | 256 × 7 = ☐ | **3** | 164 × 6 = ☐ |
| **4** | 6 × 215 = ☐ | **5** | 8 × 325 = ☐ | **6** | 7 × 284 = ☐ |

7 Draw the wheel.
Match the numbers
to be multiplied so
that all four pairs
have the same answer.

Let's investigate

Find the missing numbers
to do this multiplication.

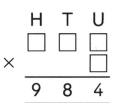

```
   H  T  U
   ☐  ☐  ☐
 ×       ☐
   9  8  4
```

Find other ways to get the same answer.

75

Number 8

Britain started to use decimal coins on 15 February 1971.
There are 100 pennies in each pound.
In 1983 the pound coin was first used.

A

 = or £1.

 is $\frac{1}{10}$ of £1.

1 is $\frac{2}{\square}$ of £1.

2 is $\frac{3}{\square}$ of £1.

3 Carry on the pattern until you reach this.

 is $\frac{10}{\square}$ of £1.

 is 10p or £0·10.

 is 20p or £0·20.

4 50p = £0·$\square$0   **5** 60p = £0·$\square$0

6 70p = £0·$\square$0   **7** 80p = £0·$\square$0

Show the coins needed to make these amounts.

8 £1·20

9 £1·40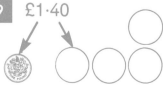

10 £1·50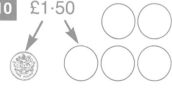

76

100 pennies = 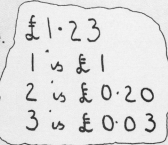 or £1

$\frac{1}{100}$ of £1 = 🪙 or 1p

> In Great Britain 100 pennies make 1 pound.
> In France 100 centimes make 1 franc.
> In Germany 100 pfennigs make 1 mark.
> In America 100 cents make 1 dollar.

11 $\frac{2}{100}$ of £1 = ☐ p

12 $\frac{5}{100}$ of £1 = ☐ p

13 $\frac{7}{100}$ of £1 = ☐ p

14 $\frac{9}{100}$ of £1 = ☐ p

🪙 is 1p or £0·01.

15 2p = £0·0☐

16 4p = £0·0☐

17 5p = £0·0☐

18 7p = £0·0☐

Draw the 1p coins needed to pay these amounts.

19 £0·09

20 £0·04

21 £0·08

Let's investigate

Use the numbers 1, 2 and 3.
Write the different amounts of money you can make.

£☐·☐☐

Write the value in pounds of each number
you use.

£1·23
1 is £1
2 is £0·20
3 is £0·03

77

B

1 Copy and finish the chart.

Pence	Fraction of £1	Till shows
30p	$\frac{3}{10}$	£0·30
50p		
		£0·02
	$\frac{4}{10}$	
	$\frac{4}{100}$	
		£0·70
7p		

	1988	1989
Large tin of salmon	£2·98	£3·79
10 litres of petrol	£3·67	£4·17

Prices of the things we buy can change from year to year.

Write each price.
Show the pence as fractions of a pound.

2 £2·98
$\diagdown$ $\diagup$
□   □
$\frac{}{10}$ $\frac{}{100}$

3 £3·79
$\diagdown$ $\diagup$
□   □
$\frac{}{10}$ $\frac{}{100}$

4 £3·67
$\diagdown$ $\diagup$
□   □
$\frac{}{10}$ $\frac{}{100}$

5 £4·17
$\diagdown$ $\diagup$
□   □
$\frac{}{10}$ $\frac{}{100}$

Look at these amounts.
How much is the 2 worth in each one.

6 £2·75   **7** £5·72   **8** £7·25

Let's investigate

Explain what the machine is doing.
Put some more amounts of money in
and show what comes out.

£2·34

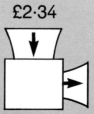

£2 + £$\frac{3}{10}$ + £$\frac{4}{100}$

C Mum bought wood to make a shelf. She cut 2·46 m.

Metric measures started to be used from 1975 in Britain, but some old measures are still used as well. For example, some people measure in miles, feet and inches as well as kilometres, metres and centimetres.

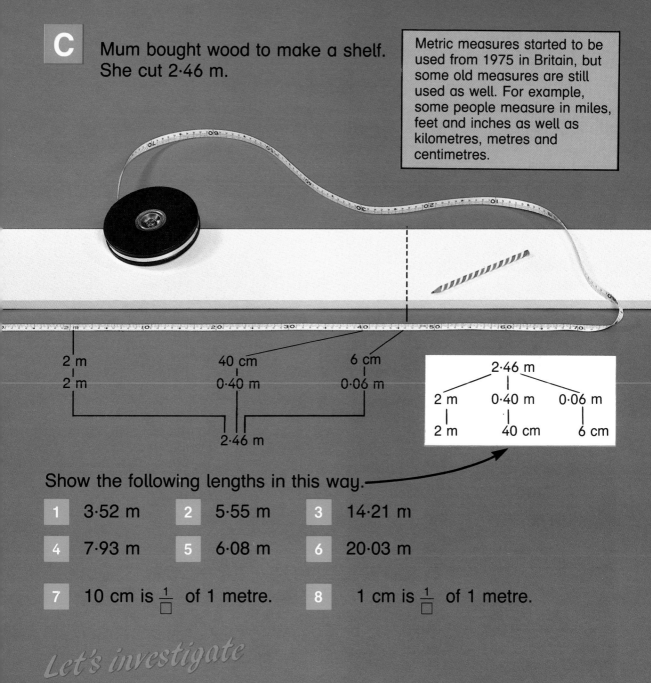

Show the following lengths in this way.

| 1 | 3·52 m | 2 | 5·55 m | 3 | 14·21 m |
| 4 | 7·93 m | 5 | 6·08 m | 6 | 20·03 m |

7 10 cm is $\frac{1}{\square}$ of 1 metre.   **8** 1 cm is $\frac{1}{\square}$ of 1 metre.

Let's investigate

Measure some objects in the classroom. Write their lengths in metres. Show how your measurements can be split into fractions.

3·67 m = 3·00 m + 0·60 m + 0·07 m

= 3 m + $\frac{6}{10}$ m + $\frac{7}{100}$ m

Explore kg and g in the same way. Record and explain your findings.

Data 2

A   Mittens had 3 litters of kittens.

> **Mittens**
> 1st litter 4
> 2nd litter 6
> 3rd litter 5

1   How many kittens altogether?

2   If all the litters were the same size, how many kittens in each one?

This is the average.

3   The average size of the litters is ☐ .

Find the average size of these litters.

4   **Dog**
1st litter 8
2nd litter 4
3rd litter 3

5   **Hamster**
1st litter 12
2nd litter 12
3rd litter 9

6   **Squirrel**
1st litter 2
2nd litter 4
3rd litter 6
4th litter 4

> The range is the difference between the largest and smallest numbers in a set.

Look at the litters of kittens.

7   How many in the largest? ☐
How many in the smallest? ☐
What is the range? ☐

Find the range of the litters of these animals.

8   The dog   **9**   The hamster   **10**   The squirrel

Another name for average is mean.
What is the average or mean for each set of numbers?

11 5, 6, 7   **12** 9, 9, 2, 8   **13** 10, 12   **14** 2, 3, 4, 5, 6

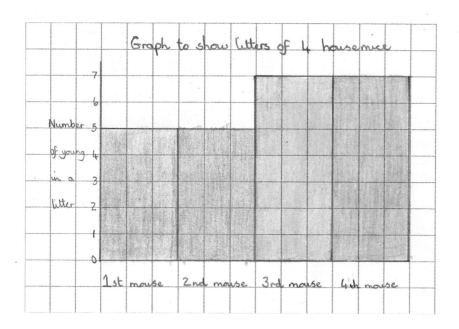

Graph to show litters of 4 house mice

Number of young in a litter

1st mouse   2nd mouse   3rd mouse   4th mouse

15 How many young mice altogether?

16 What is the average size of a litter?

17 What is the range of the litters?

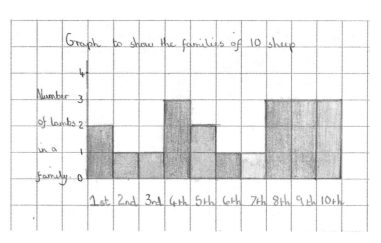

Graph to show the families of 10 sheep

Number of lambs in a family.

1st 2nd 3rd 4th 5th 6th 7th 8th 9th 10th

18 How many lambs altogether?

19 What is the average size of a family?

20 What is the range of the families?

Let's investigate

Sheba had 4 litters of puppies.
The average number of puppies in a litter is 5.
How many puppies could be in each litter?

Find other ways to do it.

81

Find the mean and range for the clutches of eggs.

A nest of eggs
is called a clutch.

1   Agama Lizard
    18   14   21   15
    Mean ☐   Range ☐

2   Nile Crocodile
    34   45   52   41

3   Python
    26   21   35   63   30

4   Giant Tortoise
    11   10   15   13   11

5   Chameleon
    25   34   28   29

6   Green Turtle
    91   105   116

This is the nest of a teal.

Find the number of eggs in the missing clutches.

	Duck	Clutches				Average
7	Teal	9	☐	12		10
8	Shelduck	10	14	☐	11	12
9	Mallard	11	☐	10		11
10	Goldeneye	8	11	9	☐	9

11 The average clutch size is 14 eggs.
One missing number is twice as big as the other.

12   ☐   ☐

Let's investigate

Find the numbers for these clutches.
The average for the clutches must be the same.

Clutches   Average
☐ ☐ ☐   ☐
☐ ☐ ☐ ☐   ☐

Find different ways of doing it.

C

In the time of Queen Victoria it was common to have between 8 and 12 children in a family.

Queen Victoria had 9 children.

Two other famous Victorians were
Charles Dickens and Benjamin Disraeli.

Dickens was one of 8 children.
Disraeli was one of 5 children.

1   What is the average number of children for these
    three Victorian families, to the nearest whole number?

2   In 1876 one Victorian family had children
    aged 15, 13, 12, 10, 8, 6, 6, 3, 1 years.
    Find the average age of these children
    to the nearest whole number.

3   What is the average age of the children in your family,
    to the nearest whole number?

Let's investigate

Find out the average number of children for the families in your class,
to the nearest whole number.

Do the same for another class.
What do you notice?

Are the averages similar to the Victorian average?

Money 2

This school decided to make a conservation area on part of their school field. They started to collect money to pay for it.

Add these pounds that were collected.

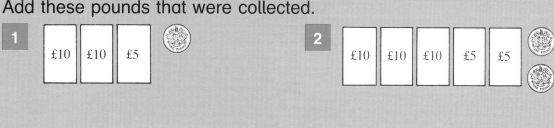

1 £10 £10 £5 (£1 coin)

2 £10 £10 £10 £5 £5 (£1 coin) (£1 coin)

3 £20 £10 £10 £5 (£1) (£1) (£1)

4 £20 £20 £10 (£1) (£1) (£1) (£1)

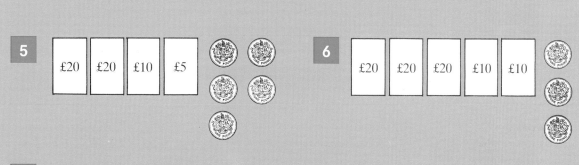

5 £20 £20 £10 £5 (£1) (£1) (£1) (£1) (£1)

6 £20 £20 £20 £10 £10 (£1) (£1) (£1)

7 How much was raised altogether?

85

As the money was collected, it was put into a bank account.
Copy the table and find the balance for each date.

8

| | money paid in | amount now in the account |
Date	Deposit	Balance
23 January	£140	£140
29 January	£100	£240
16 February	£120	
26 February	£130	
13 March	£150	

balance   £140
deposit   £100
new balance   £240

9 On 20 March the school bought some garden tools which
cost £100.
How much was left in the bank?

Let's investigate

Use a catalogue to choose some garden tools.
Spend up to £100.
Which tools would be most useful?

More money was raised at a Spring Fair.
What was the total sum collected on each stall?

1 Cakes

	£	p
notes	60	00
£1 coins	5	00
silver	2	55
bronze	1	23
total		

2 Snacks

	£	p
notes	35	00
£1 coins	3	00
silver	2	75
bronze	2	56
total		

3 Sponge throwing

	£	p
notes	20	00
£1 coins	10	00
silver	2	65
bronze	8	37
total		

4 Lucky dip

	£	p
notes	15	00
£1 coins	4	00
silver	4	80
bronze	13	43
total		

5 Golf

	£	p
notes	25	00
£1 coins	11	00
silver	1	45
bronze	9	38
total		

6 Wellie throwing

	£	p
notes	25	00
£1 coins	12	00
silver	5	85
bronze	6	89
total		

Date	Deposit	Balance
	£	£
8 June	—	200·00
26 June	120·50	320·50
17 September	100·50	
21 October	55·60	
3 November	150·25	

7 How much altogether was deposited in June and September?

8 How much altogether was deposited in October and November?

What was the balance on these dates?

9 26 June **10** 17 September **11** 21 October **12** 3 November

This chart shows withdrawals from the bank.

Date	Withdrawn	Balance
	£	£
13 November	—	735·27
16 November	115·27	620·00
30 November	200·00	

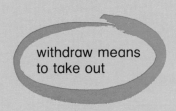

withdraw means
to take out

13 How much was withdrawn during November?

14 What was the balance on 30 November?

Let's investigate

On 10 December the balance was £420.
Two withdrawals were made both over £50.
The final balance was £310.
Find the different amounts that might have been withdrawn.

Item	Date	Deposit £	Withdrawn £	Balance £
—	8 January	—	—	450·75
—	12 January	153·27	—	
pond liner	19 January	—	100·00	
—	15 February	62·38	—	
shrubs	16 March	—	115·87	

Look at the chart. What was the balance on these dates?

1 12 January   **2** 19 January   **3** 15 February   **4** 16 March

Let's investigate

Choose some shrubs from a gardening catalogue.
Find the cost of 9 shrubs. There must be 3 different varieties.
Find the cost for different choices.

Shape 2

A

1 Use geostrips. Make these triangles.

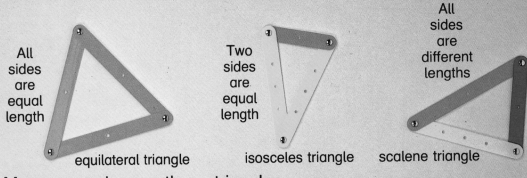

All sides are equal length
equilateral triangle

Two sides are equal length
isosceles triangle

All sides are different lengths
scalene triangle

Measure and name these triangles.

2

3

4

5

6 Draw an isosceles triangle. Use it to design a pennant.

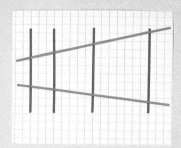

The blue strips are parallel.
The red strips are not parallel.

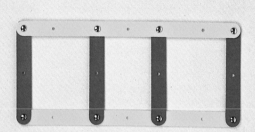

7 The yellow geostrips are _____ .

8 The blue geostrips are _____ .

9 Make a geostrip ladder.
Draw and colour it.
Use a different colour for each set of parallel lines.

10

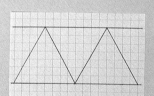

Copy this pattern on squared paper.
Colour each pair of parallel lines.
Use a different colour for each pair.

The blue straws are vertical.
The red straws are horizontal.

11 The yellow straws are _____ and _____ .

12 The green straws are _____ and _____ .

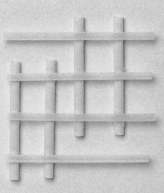

13 Make a pattern with straws.
It must have 3 vertical straws and 2 horizontal straws.
Draw your pattern on squared paper.

14 Make a chart to show some horizontal
and some vertical things you can find
in the classroom.

15 Find some things in the playground
that are vertical or horizontal.
What are they?

Horizontal	Vertical

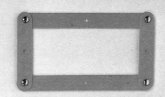

16 Use geostrips.
Make a rectangle and a square.

17 Draw them on squared paper.
Colour the pairs of parallel lines.

18 Push the geostrip square sideways.
What do you notice about the sides?

19 Draw the new shape.
Colour any pairs of parallel lines.

20 Push the rectangle sideways.
What do you notice about the sides?

Draw the new shape.
Colour pairs of parallel lines.

Let's investigate

Stop the square from moving sideways
by adding another geostrip.
Try the strip in different positions.
What shape do you make that stops the
square from moving? Draw your results.
Do the same with rectangles.

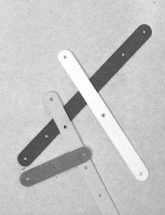

B All 4 sided shapes are quadrilaterals.
Match the labels to the quadrilaterals.

A	B	C	D
1 vertical side 1 horizontal side	No parallel sides 2 pairs of equal sides.	1 pair of parallel vertical sides	1 pair of parallel horizontal sides

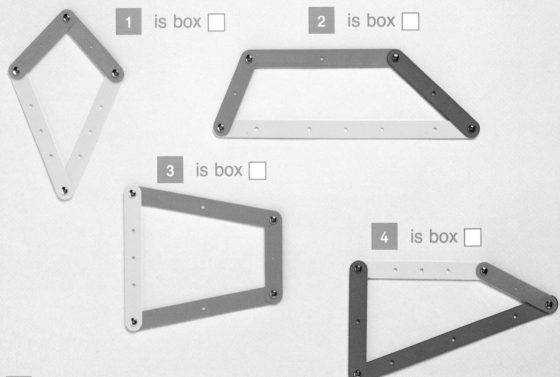

1 is box ☐   **2** is box ☐

3 is box ☐

4 is box ☐

5 Use geostrips.
Make two different quadrilaterals.
Draw the shapes.

6 Change their shapes by pushing them.
Draw the new shapes.

7 Put a diagonal geostrip on each shape.
What do you notice now?

8 What new shapes have you made?
Draw them.

9 Which is a rigid shape,
the quadrilateral or the triangle?

A rigid shape cannot be pushed out of shape.

94

gutter

drainpipe

10 Copy this house on squared paper.

11 Colour red all the lines that are
parallel to the gutter.

12 Colour blue all the lines parallel to the drainpipe.

13 Colour green all the lines that are left.
What can you say about these lines?

14 Draw a pair of parallel lines of different lengths.

15 Draw a pair of lines that are not parallel.

16 What is special about parallel lines?

Let's investigate

Draw shapes that have these.

1 pair of parallel lines
2 pairs of parallel lines
3 pairs of parallel lines
4 pairs of parallel lines

Use geostrips to help you.

C

1 Make these shapes with geostrips.

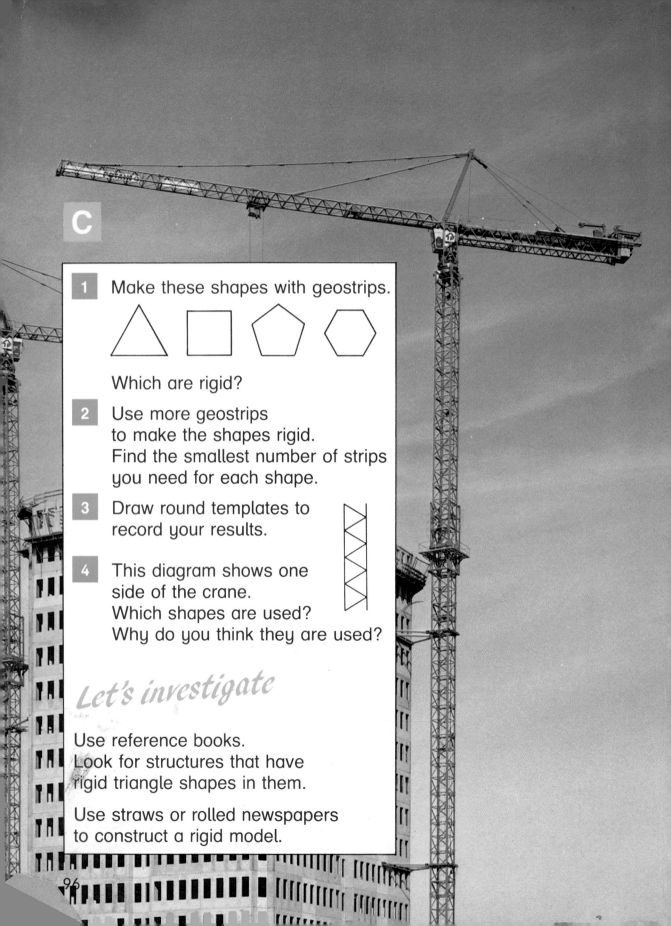

Which are rigid?

2 Use more geostrips
to make the shapes rigid.
Find the smallest number of strips
you need for each shape.

3 Draw round templates to
record your results.

4 This diagram shows one
side of the crane.
Which shapes are used?
Why do you think they are used?

Let's investigate

Use reference books.
Look for structures that have
rigid triangle shapes in them.

Use straws or rolled newspapers
to construct a rigid model.